Gr

How to Build a Photobioreactor for Growing Algae for Proteins, Lipids, Carbohydrates, Anti-Oxidants, Biofuels, Biodiesel, and Other Valuable Metabolites

by Christopher Kinkaid

Copyright © 2014 Christopher Kinkaid
All Rights Reserved
http://www.algaetoday.com

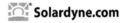
Solardyne.com

Published by Solardyne, LLC
Portland, Oregon

ISBN-13: 978-1500485450
ISBN-10: 1500485454

Table of Contents

Preface	3
About the Book	5
About the Author	8
Introduction	10
Chapter One: Growing Algae The Big Picture	13
Chapter Two: Selecting Your Algae Strain	21
Chapter Three: Build your Photobioreactor (PBR)	25
Chapter Four: Algae Optics	31
Chapter Five: Algae Nutrition	41
Chapter Six: Algae for Biofuels	47
Chapter Seven: Algae Culturing Techniques	55
Chapter Eight: FAQs about Photobioreactors	63
Chapter Nine: Quick Guide for PBR Construction	71

Preface

Algae is a miracle of Nature. Rich, in Amino acids, Proteins, Lipids, Carbohydrates, Anti-oxidants, phycobiliproteins, and other valuable products, algae is being tapped as the new feedstock across industries.

This Book describes how to build your own Photobioreactor to grow pure algae species (taxa).

Algae, are Earths "engine" to fuel the food web. As a "primary producer," responsible for nearly half the oxygen production on Earth, the power of algae is being commercialized to produce valuable organic products. Build your own, Algae Photobioreactor (PBR) grow kit, to Cultivate valuable algal strains, and tap into the rapidly growing Algae Industry.

Grow algae reliability, and repeatably, with Photobioreactor (PBR) Algae Grow Kits for controlled photosynthesis. Grow up to Four different Algal taxa using these 4-vessel Algae grow kits rated at 80 Liter total capacity.

Complete with optical, mechanical, electrical, pneumatic, and biological systems, photobioreactors give you complete control. Growing monocultures of algae, using photobioreactors, is useful for researchers, developers, companies, universities, and those who need to cultivate Algal monocultures with purity, and minimal cost of construction.

Algae, produce valuable amino-acids, proteins, carbohydrates, and essential oils (lipids) consuming water-borne pollution for nutrients. Algae species, grown with your PBR algae grow kits, enable researchers to tap algae's enormous productivity, able to double in mass in 24 hours under exponential growth phase. Algal researchers, work to develop protocols for increased production.

Growing algae converts water, in-organic compounds (CO_2), and solar radiation into valuable organic molecules. This Book is written as a resource for building your own photobioreactor, and growing valuable algal strains.

This Book is written, as a resource for researchers, to construct an effective bioreactor, rated at 80 Liters, for growing algae monocultures. Isolated from contamination, these photobioreactors, offer the researcher total control of all inputs, and thermodynamic conditions, to grow a specific monoculture algal strain.

Grow Algae for Profit, using photobioreactors, to produce useful quantities of pure species (taxa). Grow Algal Biomass, for your experiments, or for sale, with this easy-to-build Photobioreactor.

About the Book

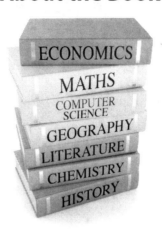

This Book is written, as a resource for building your own Photobioreactor (PBR), for Algae Growth, and Cultivation.

Your photobioreactor can be built with lab equipment readily available in a Beer Brewing Store, and other vendors. Use Glass vessels, Non-toxic tubing, Pastures Curves, and other essential elements, available in local hardware stores, to build your PBR.

Chapter One looks at the big picture of growing algae. Aquatic species have particular requirements. Algae, are very robust, yet, very delicate in their preferred conditions. The Algae Cultivator can use photobioreactors (PBR), to control the growing environment.

Chapter Two reviews different algal species of interest, and of potential, and substantial value to industries ranging from cosmetics, fish, and animal feeds, to nutraceuticals, anti-oxidants, and biofuels. Includes, species list, for your consideration.

Chapter Three describes your Photobioreactor (PBR) hardware, and parts lists. The PBR contains lighting elements, mechanical frame, an air pump, and filter system, with Pasture's Curves, to halt any contamination. The PBR kit uses Glass, and 100% Non-toxic Food Grade plastic tubing for introducing air into the growth vessels.

Chapter Four covers Algae Optics. Being a "Photo-bioreactor," algae require specific "optical" conditions for optimum growth. Chapter Four discusses the various "triggers" and requirements, which stimulate algal growth rates, and products from an optical perspective.

Chapter Five discusses the nutritional requirements of algae. As an aquatic species, algae, and diatoms, are highly sensitive to dissolved elements in the water, or lack of them. Algae growth protocols, allow the researcher to build a specific "growth profile" to cultivate a selected species (taxa), and control the metabolites produced by the algae.

Chapter Six looks at Algae for Biofuel feedstock. Oil accumulating algae are highly desired. Influencing the growth cycle of your algae for biofuels, or biodiesel feedstock, allow researchers to develop protocols which maximize lipid production.

Chapter Seven examines basic culturing techniques for measuring Growth Rates, and Net Algal Biomass production. Algae, in their growth phase go through 5 essential phases. Acclimation, Compensation Point, Exponential growth phase, saturation Point, and Collapse phase. Manipulating algae, at each point in their classic growth curve, give researchers the ability to "trigger" reactions for a desired output.

Chapter Eight looks at Frequently Asked Questions about photobioreactors, construction, and operation. Review, procedures for mixing, sampling, measuring, and growing algae cultures.

Chapter Nine is a Quick Guide to your Photobioreactor construction. Step by Step assembly of your Mechanical Frame, Growth Vessels, Air Pump, filtering, and Lighting Systems.

About the Author

Christopher Kinkaid

Christopher (Toby) Kinkaid, originally from Portland, Oregon is the founder of **Solardyne.com, SolarQuote.com,** and **AlgaeToday.com,** and has worked in clean energy technology for over three decades.

Kinkaid, is the inventor of the "**Helyx**" Vertical Axis Wind Generator, the "**Mariposa**" Non-imaging solar concentrator PV module (continuous operation at Sandia National Laboratory since 1994), the **Solar Demultiplexer** optical solar concentrating lens (Dr. James/Sandia National Laboratory 1991), and the inventor of the original "Solar Power Pack" (Mother Earth News, "Littlest Utility" June/July, 2001).

Kinkaid, has been an official lecturer and presenter on clean energy technology around the world including APEC, Bangkok, Thailand, 2003, "Energy Solutions World", Tokyo, Japan, 2003, The International Biomass Conference (IBC), 2010, Minneapolis, MN, and the Algal Biomass Organization (ABO) Conference, 2010, Phoenix, AZ.

Kinkaid, has appeared in interviews on KOIN TV, KGW TV, and "Sustainable Today" produced in Oregon. Kinkaid, has served on the board of directors for the National Hydrogen Association, in Washington D.C., 1993, and the Japan Satellite Communications Company (JCNET), Fukuoka, Japan, 1994-95.

Kinkaid, served on the board of directors for Algaedyne Corporation, Preston, MN, 2010-2013. Kinkaid, presently serves as CEO of Solardyne, LLC in Portland, Oregon.

Christopher (Toby) Kinkaid is based on the West Coast, and continues his work in Solar, Wind, and Biomass applications, research, and development in Portland, Oregon.

Introduction

Algae is a force of nature. All life on earth traces back to single cell organisms. Algae, are the base of the aquatic food web, and "engines" of oxygen, and base nutrition productivity for our planet. Half of the Oxygen on earth comes from algal micro-organisms. The intense interest from industry in "Algae," is driven by incredible growth rates, capable of converting inorganic chemicals, into some of the most valuable organic molecules on earth.

This Book is written to describe how to build a photobioreactor (PBR), for growing algae, and diatoms. The Photobioreactor (PBR), described in this Book, is designed, and constructed from Glass Vessels, and other equipment, readily available from hardware stores, Beer Brewery, and laboratory supply companies. This Book includes a complete Parts List for constructing your own photobioreactors.

The Photobioreactor (PBR) allows researchers to grow all algal Taxonomic divisions:

Baciariophyta, Chrlorarchiniophyta, Chlorophyta, Cryptophyta, Cyanophyta, Dinophyta, Euglenophyta, Glaucophytoa, Haptophyta, Herokontophyta, and Rhodophyta.

Growing algae is the ultimate in "Golidlocks" syndrome. Algae, like it "just right."

Aquatic species growth rates, are driven by a specific (range of) conditions including: pH, temperature, dissolved CO_2, dissolved O_2, macro, and micro nutrients, specific metal ions, vitamins, and a Photosynthetic Active Radiation (PAR) source of light.

A photobioreactor is a controlled environment, you create, to provide the "sweet spot" of the growing algae, by controlling, and manipulating, these conditions.

The PBR described herein is based on Glass Vessels, Non-Toxic food grade tubing, Pastures Curve, to impede any pathogen entering your vessels, pneumatic air pumps, and 5 micron filters, which remove any contamination in the incoming air.

This Book describes, the photobioreactor hardware, you can build for your Laboratory, as well as discussions on Nutrients, Lighting, Oxygenation, CO_2 injection, and culturing techniques.

Grow Algae, and Diatoms, for profit. Markets for algae are growing worldwide. Specific species (taxa) are very expensive to buy from providers, often hundreds of dollars, per Liter! Building your own photobioreactor, gives you the means of growing pure Monocultures, of algal species, which provide "feedstock" for your algae experiments.

Chapter One - Growing Algae
The Big Picture

Algae, in general, are aquatic species. Single-cell "growth" engines, which consume inorganic materials, and produce organic molecules. Algae, through photosynthesis, convert segments of solar energy, trace minerals, CO_2, and water, into the amazing process of "oxygenic-photosynthesis," which drives cell growth, and reproduction, and makes life on Earth, as we know it, possible.

As an Algae Cultivator, you're trying to emulate nature, and improve on nature, by "triggering" different effects, along the "growth-cycle," with your control of thermodynamics conditions.

Photosynthesis evolved on Earth when life needed a "battery." Fragile DNA needed protection, with early Earth bombarded with UV-C ultra-violet radiation, algae evolved responses such as photosynthetic production of many organic molecules which increased the algae's survival response. Accessory pigments, the mechanisms evolved in the algae, to collect more of the available solar radiation, or the production of anti-oxidants, to "wrap" up the precious, and vulnerable DNA.

Algae reproduce during the night, probably due to the massive presence of UV bombardment during sun-hours, on ancient Earth. Replicating DNA, at night, minimized the disruption which may be caused by energetic Ultra-violet (UV) light in

Photobioreactors, such as the kit described in this eBook, provide a means for researchers to "influence" the strain being grown, by changing the environment, according to your growth protocol, for a desired result.

Stressing algae for selected production

Algae, which are "stressed" at strategic times in the growth cycle, can produced selected molecules of interest to the cultivator. Selected molecules are the objective of growing algae.

Algal biomass is produced when "energy" from photosynthesis "exceed" those energies used for cell respiration, and cell division. The specific growth rate, of your algae, will be "thermodynamically" determined by "how" you grow your algae.

The Photobioreactors (PBR), described herein, allow you to adjust the optical, temperature control, rate of CO_2, and O_2 flow into the culture, the pH, and nutrient mix, by what you add to your growth vessels, and the "timing," and "rate" at which you harvest.

Manipulating nutrients, the intensities, selection of wavelengths, and photoperiods, in your light source, temperature, pH, dissolved CO_2 and O_2 levels, will have dramatic impacts on algal metabolism.

The specific algal growth-rate is the rate-of-change, of algae biomass accumulation. The rate of "anabolic" processes (photosynthesis), and "catabolic" processes (respiration) will determine your net biomass gain.

Manipulating nutrients, levels of light (photosynthetic photon flux density, PPFD), photoperiod, temperature, pH, and dissolved O_2, and CO_2 levels, have dramatic, and controlling impacts on algal metabolism.

Photobioreactors, allow researchers to test Growth Protocols, by systematically adjusting major thermodynamic parameters, such as Temperature, Light Level, Photoperiod, as described, and is a useful tool for research, and commercialization.

Algae, often use the Primary Pigment Chlorophyll-a. Found throughout the phytoplankton kingdom,

Chrolophyll-a is, probably, the most valuable life giving molecule on Earth.

Algae, have evolved "Secondary Pigments" which tap other wavelengths in the spectrum to drive chemical processes. Other pigments, respond to additional wavelengths in teh solar spectrum, and give algae an added means of converter for energy to survive. Algae, harvest this additional solar spectrum, to gain extra energy for metabolism, respiration, and cell division.

Secondary pigments, or often called "accessory" Pigments include Chlorophyll-b, Chlorophyll-c, Carotenoids, and phycobiliproteins. Additional pigments provide an evolutionary advantage, thermodynamically, to the algae cell. Our, advantage is we can harvest the valuable "metabolites," and products which result from these additional pathways.

Secondary Pigments, provide the algae with valuable molecules, such as Anti-oxidants. High UV radiation levels, as well as chemical stresses, threatened the DNA of early algae. The protein Astaxanthin, highly valued, was evolved by the Algae to be "Sun Block" by being highly absorptive of UV light.

Algae, would produce Astaxanthin, (bright red in color), which absorbed the UV after wrapping itself around the DNA proteins to protect the valuable cargo. When chemical, or UV stress occurred to the

Algae Cell, a pathway evolved to produce Astixanthin to protect the cell.

Algae, are extremely sensitive to their condition, and changes (rates of changes) to their environments. Controlling these conditions, with your Photobioreactor, allow you influence your algae to produce specific molecules of interest.

Balance in all things

The photobioreactor, begins with a lighting system. Photo-autotrophs, are highly responsive to optical energy. The most influential aspect of growing algae is the optical regime you use in your growth protocol. The optical regime addresses wavelengths, intensities, and photoperiods.

Chlorophyll-a, responds to specific wavelengths of light, while secondary pigments respond to other wavelengths.

Photobioreactors provide a platform to grow specific species (Taxa), and develop Growth Protocols to enhance Algae's natural productivity. How you use your photobioreactor, with a schedule of actions, measurements, and harvests you choose determines the yield.

Algae produce many valuable compounds vital for cosmetics, and nutraceutical markets. The natural oils, and lipids, rich in Omega 3s, and highly valued.

The human body has evolved along with algae, and from algae. The natural oils, and anti-oxidants, are not rejected as often, compared to synthetic products for consumers.

"Haematococcus pluvialis", a Chlorophyceae (Green Algae), produces more anti-oxidant Astaxanthin, about 40,000 ppm when "stressed," than any known organism on Earth. This makes (H. p.), very valuable for the nutraceutical, and cosmetic markets.

Natural Astaxanthin, has a market value, in the thousands of dollars, per pound, and ihighly valued in the nutraceutical, and aquaculture markets.

Algae, have incredible mechanisms to enhance the production of photosynthetic products when "stressed." Algae biomass growth, has nutritional and other need you can manipulate during their algal growth cycle to produce desired organic products. Stressing algae, increases, or decreases, something the algae needs during its life-cycle.

Stressing, is changing the environment of the algae, to produce a predicted response, such as the production of **Astaxanthin**.

The Photobioreactor Kit, described below, provides the equipment you need to grow, and influence, your algae's growth profile.

Algae have highly responsive "metabolisms" you can influenced to produce higher levels of selected organic products, including amino-acids, proteins, organic-dyes, anti-oxidants, vitamins, and important for biofuel production: lipids.

Lipids (oils), are the principle raw-material for biodiesel (both vegetable, and animal-based fatty-acids can be used as a feedstock). Fatty-acids, can be transesterified into biodiesel.

The lipids produced by Algae, are often categorized as "Storage" lipids (non-polar), and "Structural" lipids (polar lipids). "Storage" type lipids, large Triaclglycerides (TAGS), can be transesterified to produced biodiesel.

Researchers, have investigated influencing algal cells for biodiesel production by "limiting" some variable in the growth cycle. "Tricking" the algae, by changing some condition, can induce the production of a desired "molecule" production, as part of the biomass produced. Photobioreactors (PBR), allow the algal cultivator to adjust conditions, such as temperature, pH, light levels, presence, or lack of chemical nutrient to produce a desired output.

All life on Earth, with few exceptions, depend on oxygenic-photosynthesis, as the primary process which produces nutrition (for the base of the food chain), and oxygen.

Photosynthesis, is the "primary producer" of all nutrition, and oxygen - upon which life on Earth, and the oceans depend. The "power supply" for photosynthesis, is the sun, delivering a peak power at the surface of the Earth of 1,000 watts/square meter.

To stimulate photosynthesis, you'll need to produce the wavelengths which dominate the response characteristics of the algae's Primary, and Secondary pigments. Each algae, will have their particular sweet spot of all thermodynamic factors.

Chapter Two - Selecting your Algae Strain

Buying monocultures (pure species), of Algae is expensive - often hundreds of dollars Per Liter!

Photo-bioreactors can be used to grow algal monocultures, and save, over time, potentially, thousands of dollars in algal culturing costs.

Algae species of interest, are selected because you want a specific, or multiple molecules of value. Selecting algae, is to work the problem "Backwards." Start, with what you want to end up with, after Growth. The species (taxa) you choose, depends on what you want to produce as an end product. Are you looking for oils, (lipids), for the biodiesel, or cosmetics industry? Are you looking for complete proteins (essential amino-acids), for the fish feed market?

Your choice of algae depends on your goals. The following list of algae, for example, are listed with a range of lipid content (dry-weight). Each species (taxa) has its own growth protocol, and growth rates. Lipid content of your Algae batches depend on your cultivation technique, how you inoculate, and start your culture, the growth-media you add to your Glass Growth vessels, the light regime you apply, and how well you control pH, and temperature.

Pure monoculture species are highly valued and sell for hundreds of dollars per liter. The following is a list of useful, and valuable Algal Species (taxa):

Chlorella vulgaris

Chlorella minotissima

Ankistrodesmus sp.

Crypthecodinium cohnii

Scenedesmus sp.

Cyclotella sp.

Dunaliella tertiolecta

Hantzchia sp.

Nannochloropsis

Neochloris oleoabundans

Nitzschia sp.

Phaeodactylum tricornutum

Stichococcus sp.

Nannochloris

Thalassiosira pseudonana

Tetraselmis suecica

Botryococcus branuii

The superstar Chlorella vulgaris - is well studied for its high productivity. Algae biodiesel, based on Chlorella vulgaris has advantages to offer in terms of high growth rates, and some issues to be addressed, including the fairly hard cellulosic cell wall, which needs to be "broken" to reach the internal oils.

Chlorella vulgaris, a Chlorophyceae, grows well using well known ratios of nutrients C:N:P:K. Limiting Nitrogen (ratio to other nutrients), and Chlorella vulgaris responds, producing more starches, and polyunsaturated fatty-acid lipids. Polyunsaturated fatty-acids, are a great prize. The "nutrient-limited" algae sense a small crisis, and produce more lipids to "store up" energy for an anticipated deficit.

If you're choosing an algae strain for poly-unsaturated fatty-acid lipid production, Chlorella vulgaris is a good choice. Chlorella minotissima, from the Phylum Chlorophyta, when Nitrogen-limited produced 39% EPA (omega-3 fatty-acid Eicosapentaenoic-acid, highly valued in the nutraceutical, and biodiesel markets.

Nannochloropsis, demonstrates great algae biodiesel production when "influenced" by nutrient limitation. Nannochloropsis, is composed of six identified taxa, each promising, and live in salt-water, fresh water, and brackish water.

Nannochloropsis, grown under the right conditions, can accumulate up to 60% dry-weight of polyunsaturated fatty-acids, in Nitrogen-limited growth protocols. This makes Nannochloropsis highly valued, as a potential feedstock, in the biofuels industry.

Chapter Three - Build your Own Photobioreactor

You can build your PBR using 4 Glass Growing Vessels. You'll build a PVC frame, and place two Florescent lights on top of the PVC frame above the growing vessels. You'll place aquarium pumps to pump air, and CO2 into the vessels. The Vessels have "Stoppers" on the top 2 hole type.

Your Photo-Bioreactor system will Include:

Heavy Duty Timer, Mechanical Frame, made from PVC piping from hardware store.

Four (4) 20 Liter Growth Vessels. Glassware Growth Vessels with 100% Non-Toxic Food-Grade tubing, plugs, and fittings.

Pneumatic Air Pumps with Inline Bacterial Filters for sterile aeration and mixing, with "Pasteur's Curve" exhaust vents, to prevent taxa contamination

Easy to assemble and sanitize for different taxa production runs.

Rated at 80 Liters, the four vessel PBR (20 Liters each), can be used, all for one monoculture of algal taxa. You can also use each vessel to grow completely different and separate taxa - up to Four Different Taxa with this Algae Growing kit.

Each glass growing vessel is independent of the other vessels, with their own Bacterial filters and "Pasteur's Curve" exhaust vents.

Complete Photobioreactor (PBR) Kit includes:

Mechanical Elements
Pneumatic Elements
Biological Filter Elements
Optical Elements
Electrical Fusing/Photoperiod Timer System

Biological filters, for Each vessel, sterilize air flow into your culture vessels, and "Pasteur's Curve" output vents, disallow contamination to back-flow into your culture vessels.

Use Glassware, Pyrex glass tubing and 100% Non-toxic Food-Grade material for sensitive components.

Complete Optical System produces PAR light with over 200 micro-moles/m2/sec photon flux density, adjustable by changing lamp heights, and includes with Heavy Duty Timer. Kits also include all Glassware and fittings, Pneumatic Air Pumps, Mechanical Frame, Fused Electrical system - Everything you need (hardware) to start growing algal cultures.

All PBR Algae Grow kits include Non-Toxic Evaporative Sanitizer for repeated cultivation DIY Photobioreactor PBR Algae Grow Kit Includes:

Two (2) High-Efficiency T8 Lamp Ballasted Florescent Light Fixtures, Four (4) 6500K High-Efficiency Lamps (20,000 hours). One (1) Heavy Duty Timer (plug the lamps into the timer to set your photoperiod).

One (1) Heavy Duty Fused Power Strip.

One (1) Mechanical Frame Kit. Precut and Fitted for Easy Assembly. The "mechanical" frame is composed of PVC piping (3/4" to 1.5", you choose) available at the hardware store. Cut pieces as follows:

Eight (8) Length Segments (18" each)
Eight (8) Side Segments (22" each)
Six (6) Vertical Segments (20" each)
Eight (8) 3-Way Corner Connectors
Eight (8) 3-Way Middle Connectors

Assembles into Framing as Shown Above. The Frame supports the Lights, and "defines" an interior space where the growing vessels are placed, under the lamps.

Four (4) Glass Growing Vessels rated at 20 Liters Capacity Each.

Four (4) Pyrex Glass Tubes for aeration input into the growth cultures.

Four (4) 100% Non-Toxic Food-Grade Growth Vessel Top Seals/Tubing/Fitting sets

Four (4) 100% Non-Toxic Food-Grade "Pasteur's Curve" exhaust Vents

Two (2) High-Efficiency Air Pumps (4000 cc/minute) over Four Growth Vessels. Attach a "splitter" so you can aerate 2 vessels, independently, from one air pump.

Four (4) Check-Valves (to protect the air pumps)

Four (4) Inline Bacterial Filters (one for each Growth Vessel) rated at 0.22 um. Place the inline Bacterial Filters between the air pump, and each Growth Vessel.

Twenty Two feet (22') 100% Non-Toxic Food-Grade Tubing Line

One (1) Liter Evaporative Sanitizer Solution 100% Non-Toxic Food-Grade

Total Kit contains (96) Parts.

Power Rating: 148 Watts

Cost to Operate: Less than 2 Cents per Hour (at 12 Cents/kWh electricity rates)
Footprint: 8 Square Feet, Height: 3 feet, Width: 2 feet, Length: 4 feet, Weight: 57 lb.

Chapter Four - Algae Optics

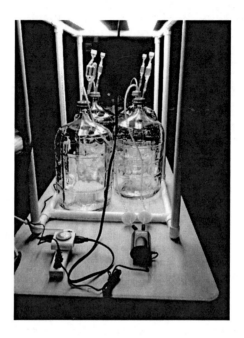

Photon wavelengths, intensities, and photoperiods are crucial as algae need a "Goldilocks" condition to reach exponential growth.

Add too-much light energy and you'll induce "light saturation." Light saturation, is when you've overloaded the Photo-reaction centers in the cells, and no more light can drive the process. Indeed, if you reach "light saturation" conditions then you'll inhibit photosynthesis, this effect is Light-inhibition.

Add too-little photon intensity, and you won't reach the "compensation" point required for net photosynthesis. Compensation is when your algae produce a net gain in algal biomass. This "Compensation point" is where photosynthesis exceeds the energy required for respiration and cell division.

Algae grow when photon intensity is between the "compensation point" and the "light saturation" point in the growth curve. Note: one of the biggest mistakes made by algae growers is to use too-much light.

Thermodynamically, once you reach "saturation" levels with light intensity - no additional photons added to the system will drive the process and further or faster. Adjust the height of your frame to adjust the intensity of light.

Use a Quantum Meter when possible, to accurately measure Photosynthetic Active Radiation (PAR) from 400nm to 700 nm, power density in micro-moles photons (micro-Einsteins)/m2/second. Photo-periods are vital to growing algae. The diurnal day-night cycle is a fundamental influence on how algae evolved.

Your choice of Photoperiods have dramatic impacts on the life-cycle of algae, as each species has its preferred day-night cycle.

LED technology is allowing researchers to match the "emissivity" of LED emitters to the "absorptivity" of primary and secondary pigments in Algae.

However, LED's often don't exactly match the wavelength "peak" response of some pigments.

New Organic LEDs (OLED) will allow LED emissivity to be "tunable" and lock into exact pigment peak wavelengths. The adoption of LED emitters to growing algae will provide a high-efficiency (you're only energizing wavelengths you need), low-temperature (LEDs run cool), and high control over intensity and duration.

Photobioreactor Algae Grow Kits use T-8 type lamps you can use any number of selected lamps that meet this physical format. LED lamps for T-8 fittings can be sourced online, or locally.

Use photobioreactor algae grow kits to grow algae for biofuels, and biodiesel production. Biodiesel production, using algae, has tremendous market opportunity as major transportation industries put pressure on diesel producers to use more biodiesel.

The biodiesel market is hugh including-trucking, trains, vessels, farm equipment, construction equipment, not to mention, many cars and trucks can run on biodiesel. Biodiesel from algae using water waste-streams, overloaded with excess Phosphorous and Nitrogen can be tapped as nutrients. Nutrients are highly valued, especially phosphorous.

Biodiesel from algae, used to clean water streams of valuable P:K:N combine to produce two revenue streams: income for cleaning the environment, and income for biodiesel produced.

Algae have "Accessory pigments" such as Chlorophyll-b, and Chlorophyll-c, absorb at peaks which are both in the violet-blue and orange-red bands but slightly varied. Other accessory pigments including carotenoids (beta-carotene), absorb in shifted wavelength peaks to capture wavelengths different than the primary pigment chlorophyll-a.

In the case of Chlorophyll-a, and Chlorophyll-b each "peak" must be activated simultaneously. Each - together - drives a photochemical pathway active in photosystems II, and photosystem I which drive the Light-Dependent processes in photosynthesis.

Photosynthesis operates in two separate parts: light-dependent reactions (in the photo-reaction centers) "oxidize" water, and the light-independent reactions (Calvin-Benson cycle) which "reduce" CO_2 to produce the building blocks of all other organic molecules: simple sugars.

Photobioreactor Algae Grow Kits are designed for the algae researcher. Grow algae for biodiesel and nutraceutical projects. Develop specific pH, temperature, lighting intensity, lighting photo-periods, nutrient recipes, and other variables to maximize your algae outputs.

During photosynthesis algae "oxidize" water to harvest an electron and a proton - releasing Oxygen as algae's waste product.

Water is oxidized producing one electron and proton pair. Once formed, the charge particles are separated, creating e a "potential difference" to drive an electron transfer chain transporting that charge to be used later by the Calvin-Benson cycle to build organic molecules.

The Calvin-Benson cycle chemically "reduces" CO_2 (carbon fixing) and builds simple carbohydrates to store energy.

Algae have Optical needs. Photo flux density, the rate of energy delivery to your algae culture, has been measure over a wide range from as little as 2 micro-moles of photons/m2/second to a more usual 80-200 micro-moles photons/m2/sec.

Photon energy for growing algae in photo-bioreactors has three major considerations:

Photosynthetic Wavelengths
Photon Intensities
Photoperiods

Photobioreactor algae grow kits provide control over all three optical factors.
Using T-8 universal lamps you can energize different spectra lamps in your included lamp fixtures to

design all kinds of optical algae growth experiments.

Photobioreactor algae growth protocols give you the control over light penetration. Ponds, and other outdoor algae growth approaches faces a big problem with "light inhibition."

Light Inhibition occurs when algae grow on the surface of a pond, and block light from penetrating the water column. This surface algae growth "shades" the algae underneath and produces an inhibition of growth.

A paradox to growing algae in ponds is the more you grow, the more algae are shaded. Light inhibition limits the productivity of ponds to a depth of about 1 - 2 cm. Algae are aquatic species which require specific environmental conditions to grow. These include Temperature, pH, dissolved CO_2, dissolved O_2, nutrients available, macro and micro, PAR light between 400-700 nm and a regular photoperiod.

Photosynthetic Photon Flux Density (PPFD) describes the energy delivery of your optical system. Power-densities for PAR light required by a specific taxa range by species from as little as 2 micro-moles photons/m2/second, for Arctic algae to over 200 micro-moles photons/m2/second for more typical algae species.

PBR kits are designed to produce a nominal 300 micro-moles/m2/sec of PAR light. You can vary this amount by adjusting the height of your lighting system.

Algae Grow Kits include complete Framing, Lighting System, Power control system, Glass and Pyrex glass growing vessels, bacterial filters, "Pasteur's Curves" and Air Pumping system. Photobioreactor Kits are designed for you to grow algal monocultures of valuable algae.

Algae are equipped with chloroplasts (containing the photoreaction centers) and occur all over the cell surface. Light entering a water column is either absorbed, or diffracted as it travels. Particles, in the water including algae, scatter the light which is not absorbed. Scattering light is an advantage to algae as it "normalizes" the photons direction and allows the cells to capture and utilize photons from all directions.

Photons in water will "scatter" and "absorb" in all directions including up again, giving light in water a very active profile of up welling and down welling normalizing photon paths - evening out the light (photon) distribution within the algae water column

Photo-periods are vital to growing algae. The diurnal day-night cycle is a fundamental influence on how algae evolved. Photoperiods have dramatic impacts on the life-cycle of algae and each species has its preferred day-night cycle.

Most algae growth occurs using a 12-hour on, and 12-hour off photoperiod. However, lengthening, or shortening the ratio has impacts on cell physiology and response. If "sun-hours" are increasing, the alga know that summer is coming and increases photosynthetic response.

If you shorten "sun-hours," alga respond to "coming winter" producing more lipids. Algae for Biodiesel offers a real source of Carbon-neutral transportation fuels. Algae for Biodiesel can be grown using wastewater streams from Agricultural, and livestock sources, with true Carbon neutrality. Carbon for algae growth comes from the atmosphere, and is returned to the atmosphere when consumed.

Photosynthetic Pigments are proteins available to capture specific photon energy vital for photosynthesis.

Light, (photon energy), is the most important factor to consider for growing algae. (Although all thermodynamic conditions are important). Photosynthesis is the primary mechanism driving algae growth and its importance to commercially growing algae is dominant. Algae require specific wavelengths of photon energy ranging from 400 nm to 700 nm.

Photosynthetic Active Radiation (PAR) light refers to the broad spectrum of wavelengths to which pigments May respond. All oxygenic photosynthesis on Earth is driven by wavelengths between 400 nm and 700 nm - not even one "octave" of light frequencies - a very narrow band given the wide electromagnetic spectrum.

The Primary pigment used throughout the algae universe is Chlorophyll-a. Chlorophyll-a, perhaps, is the most important molecule on the planet, because of its ability to capture those photons needed by the Photoreaction Centers II, and I, to drive photosynthesis light-dependent reactions.

Secondary Pigments such as Chlorophyll-b, Chlorophyll-c, carotenoids, and phycobiliproteins are proteins that capture and absorb selected photons. Energizing a "cascade" of reactions this photonic capture function is most important. Maximize your algae's pigments by stimulating both "peaks" in their absorptivity spectra.

Photobioreactor Algae Growth Kits allow you to control the optical conditions such as intensity of Photosynthetically Active Radiation (PAR) light vital in growing algae.
Photosynthesis for algae operates over a wide range of conditions depending on the species, but the wavelengths, and intensities of photon energy are, thermodynamically, most important.

Algae grow using PAR light ranging from 400 nm through 700 nm wavelengths. Intensities of PAR light vary from as little as 2 micro-moles photons/m2/sec for arctic algae to 200 micro-moles photons/

m2/sec for more common algal species. Each species will have a preferred photon intensity, collection of active wavelengths, and photoperiod to allow a light and dark cycle.

The exact wavelengths algae can utilize in oxygenic photosynthesis depend on the primary pigment (chlorophyll-a) which has two absorption peaks, one in the violet-blue and peaks again in the orange-red part of the spectrum.

Chapter Five - Algae Nutrition

Growing algae depends on many factors including the nutrient growth media you choose for your specific species (taxa).

Limiting nutrients, such as Nitrogen, has an effect in many algal species to produce more lipids. Researchers use these nutrient and other factor limitations to stimulate the algae to produce a desired organic product. Chlorella vulgaris, is well know to produce significantly more lipids, and starches when Nitrogen-limited.

Photobioreactor (PBR) Algae Grow Kits provide a tool for algae researchers to design specific nutrient mixes which enhance growth-rates, and net algal biomass production.

Algae, diatoms, and cyanobacteria require macro and micro nutrients, dissolved ions, trace metals, and several vitamins to thrive. Growth media for algae are grouped depending on freshwater, or salt water species. There is no universal growth media recipe that works for all taxa, so researches are forced to give great care to how growth media is composed, stored, and used.

Algae growth media recipes

Macro-nutrients required by algae, diatoms, and cyanobacteria include Carbon, Nitrogen, Phosphorous, Silicon, and major ions including Na, K, Mg, Ca, Cl, and SO4 as a base media.

Micro-nutrients are trace amounts of essential elements, and these include iron, manganese, zinc, cobalt, copper, molybdenum, and a small amount of metalloid selenium.

Vitamins - are vital to algae growth, specifically three: vitamin B1 (Thiamini - HCL), vitamin B12 (Cyanocolbalamin), and vitamin H (biotin). Many algae only need one or two, depending on species, but there seems to be no harm in using all three.

The addition of Trace elements are a delicate business in algae cultivation. Just small amounts of trace metals, such as iron, copper, zinc, and cobalt, are essential for photosynthetic processes. Note: all trace elements are toxic to algae if concentrations are too high. Great care should be taken not to mix up micro-grams/Liter with milli-grams/Liter.

The element Iron - is needed by all phytoplankton as it serves essential metabolic functions in electron-transport.

The element Manganese - is an essential component of the water-oxidizing centers of photosynthesis.

The element Zinc - like Manganese, is used by algae, diatoms and cyanobacteria for a variety of metabolic functions. One major use of zinc is in the formation of "carbonic Anhydrase" - this essential enzyme is vital to CO_2 transport and Carbon-Fixing.

The essential element Copper - is vital for all phytoplankton life because of its function in "cytochrome oxidase", - an essential protein in algae cell respiratory electron-transport.

Growth Media Nutrient recipes are as well guarded as those of a Master-chef in the culinary arts.

Develop your own recipes and discover the perfect blend of nutrients to drive algae's exponential growth.

Freshwater algae species typically use growth media divided into three broad categories: synthetic, enriched, and soil-water. Synthetic growth media is "artificial" media designed by the algae researcher to provide a simplified and specifically defined media. Examples include

"Bold's Basal Medium," Chu #10 medium, BG-11 medium, and WC medium.

There is a great art in preparing freshwater growth media for algae - be sure Not to use tap or distilled water. Trace metal contaminates in tap and distilled water can poison your algae.
Enriched growth media is prepared by adding nutrients to natural stream, or lake waters, or by enriching a "synthetic" media with soil or plant extracts. Enriched media is not defined, because of the unknown organic and in-organic compounds present.

Algae Pioneer Redfield (1938) describes methods for keeping continuous isolated cultures of marine diatoms - rich in Omega 3 oils - in large quantities for his laboratory experiments.

Redfield's procedure included strategically harvesting algal biomass at a certain point in its exponential growth-phase. Kilogram quantities of dry-mass diatoms were cultivated, and harvested, for his laboratory and aquaculture experiments.

Redfield is famous in biology, for the "Redfield's ratio" of photosynthetic composition vital for nutrient mix recipes, used to grow Algae. Redfield's ratio of 106 Carbon : 16 Nitrogen : 1 Phosphorous is a cornerstone of algae growth protocols and has been modified, by many researchers, to include Trace metal-ions that are required for dynamic algae growth.

Photobioreactor Algae Grow Kits provide a tool for measuring algal biomass growth rates, and growth amounts through growing algae directly.

Growing algae requires management, planning, and execution of a specific growth protocol.
Algae species have very specific appetites for growth media, and there is not universal nutrient mix which works for all species equally. Therefore, researchers use photobioreactors to control photosynthetic growth within a controllable environment.

Using Soil-water media is method of enriching a growth media using natural elements found in soil. Choose as "clean" a soil as possible. Do not choose clay, and dry the material over low heat.

When dry, run through a sieve and sift the soil into small particles. Add to water and let settle to the bottom. Natural diffusion will allow essential humic compounds and characteristics in the soil including pH, conductivity, organic buffers, nutrients and vitamins to diffuse into your growth media.

Photobioreactor PBR algae grow kits allow you to experiment with nutrient protocols and grow more algae. Develop your own nutrient recipes for the specific algae you want to grow.
Water-quality is the most important starting point when designing your nutrient growth media. dH2O generally refers to distilled, or deionized water.

Don't use dH2O (distilled) water because of trace-ion contaminants present.

Use Reverse Osmosis water, or Glass distilled water for a starting point for your synthetic media recipes. Nutrient mixes are composed then autoclaved to sterilize the water before you introduce algae inoculant.

Chapter Six - Algae for Biofuels

"The use of vegetable oils for engine fuels may seem insignificant today, but such oils may become, in the course of time, as important as petroleum, and coal-tar products of the present time." (Rudolf Diesel - 1912)

The liquid-fuels market, in the US alone, exceed $1.8 Billion per day. Crack the algae oil-accumulating protocols, and those markets can be tapped with Carbon-neutral Algae based fuels.

Oil-accumulating Algae, and Diatoms, are the key to large-scale Algae-based Biofuels and Biodiesel markets.

Diatoms and Algae can be grown with photobioreactors. Algae, as the primary feedstock for biofuels and biodiesel, is achieved with productive oil-accumulating algae, in their dormant, or resting state. Use photobioreactor PBR kits to grow algae, and conduct your own experiments to increase algae bio-products.

Growing algae for biodiesel represents the greatest market opportunity of the 21st century. Transportation fuels including biodiesel represent a multi-billion dollar market - daily. Algae biodiesel meeting this demand requires over 80 Million barrels of vegetable oil production daily. Algae for biodiesel can produce this high-volume because our organic waste-streams far exceed this value.

Algae for biodiesel makes a strong economic argument, as water pollution waste-streams contain the most of the nutrients needed to grow algae at large scale. Water ways are over-stressed with Nitrogen, Phosphorous, Potassium and other elements in our water pollution sources. Algae for biodisel can clean, (Carbon neutral), and "treat" this water pollution producing cleaner water and biodiesel fuel. Water pollution can be "redirected' to grow algae for biodiesel production solving two problems simultaneously.

Organic waste-steams currently "dumped" into fragile water-ways can be tapped as the principal nutrient source for growing algae biodiesel. Algae biodiesel can be produced in many locations using local organic waste-streams increasing energy security for those algae-based biodiesel networks.

You can select algal species for their Lipid output for Biodiesel feedstock. If your interest is Ethanol, then you're looking for a particularly Starchy strain.

Growing algae for biodiesel production starts with specific algae biodiesel growth protocols

Photobioreactor (PBR) Algae Grow Kits are designed for growing algae under your growth protocols to produce organic molecules of interest. Algae based biodiesel seeks to take water pollution resources (N,P,K) and redirect them as a feedstock to produce algae for biodiesel.
Photobioreactor algae grow kits allow you to vary the major thermodynamic parameters. Control light intensity, wavelengths, photo-periods, growth nutrient media, pneumatic aeration, and algal species.

Many algae growing techniques exist, and have been described to "nudge" algae to produce more of what you want. Algae for biodiesel looks for unsaturated lipid production - most efficiently transesterified for producing biodiesel.

Select your algae species based on the lipids you wish to produce. Select your algae on the nutrients you intend to use. Algae biodiesel requires you work out how to seed, grow, nutrient manage, harvest, dewater, and dry your algae in a commercial process.

Choose your species of algae for biodiesel on how you, or others intend to separate the oil from the algal biomass. Many companies and universities are developing oil separation techniques you can access. The most common is a centrifuge.

Algae for biodiesel requires commercially scalable technologies, and all start in the laboratory growing algae with photobioreactors.

Growing algae for biodiesel requires all inputs and processes to be quantified and repeatable. Work out your nutrient regime and photo regime as you develop your growth protocols.
Nutrient limiting, temperature variation, light level variations and photoperiods, pH, and other "stresses" can trigger algae response.

Nitrogen limiting has been often reported for "inducing" algae to produce more lipids.

Algae for biodiesel is an "engine" of fast-growing biomass that can be tapped for oils. Algae biodiesel production has many value streams. Growing algae for biodiesel production seeks to "influence" algae to produce more oils.

Oil production in algae can be "induced" with variations of their input requirements producing more polyunsaturated fatty-acids consuming water pollution in the process.

Algae for transportation fuels is an important part of the great transition of the 21st century into a sustainable industrial society.

Use photobioreactor algae grow kits to grow and research algae for biodiesel production. Algae for biodiesel is usually "processed" first by removing the oils from the algal biomass. The "press-cake" solids remaining are a great food for animals and fish farms.

Algae Press-cake with most of the oils removed for biodiesel leaves the algae less oily - ideal for nutrition management. The "oils" have been removed making the "press cake" more suitable for animal and fish food.

The "press-cake" is rich in amino-acids, essential proteins, anti-oxidants, vitamins and trace oils excellent as animal and fish feed. The removed oils are then processed through transesterification to produce slippery and stable algae biodiesel.

Algae biodiesel technology Cleans Water, Produces valuable Animal and Fish Feed, and produces Algae Biodiesel for use in diesel engines for transportation and power-production markets.

Algae offer big opportunities for producing oils (lipids) because of their high inherent-efficiency, and their ability to use waste-products as nutrients.

Researchers, and companies, have better understanding about how to provide and control growth environments, such as with AlgaeToday's Photobioreactor PBR Algae Grow Kits, to grow algal monocultures which produce high levels of valuable - selected - organic compounds of great value to industry.

For Algae-based biofuels and algae biodiesel growing oil rich algal strains that accumulate lipids are key.

One of the great pioneers in growing algae, and researching photosynthesis, was Otto Warburg (1919), in Berlin, Germany. Warburg succeeded in growing dense cultures of Chlorella, and many other species (taxa). Warburg was a great visionary for using algae as feedstock for animal, fish feeds, and biofuels.

Algae biodiesel offers many advantages for the transportation markets. Available everywhere - organic wastes as a nutrient feedstock enables algae biodiesel production in all countries.

Algae for biofuels uses the powerful engine of photosynthesis to do industrially what plants do naturally: recycle Carbon.

Algae biodiesel is carbon-neutral. Carbon-dioxide CO_2 in the atmosphere is captured and converted into proteins, carbohydrates and Lipids (oils) by Carbon-fixing using Chlorophyll-a, and other pigments driving photosynthesis. Carbon is "reduced" as water is "oxidized" fixing carbon into molecules of life.

Algae biodiesel uses the lipids for transesterification into stable biodiesel.

Consuming, or burning algae biomass oxidizes the organic compounds reforming CO_2 which returns to the atmosphere. The Carbon Cycle for Algae is Carbon Neutral - No new CO_2.

The transportation fuels market in the US alone is over $1.8 billion dollar per day. Algae for biodiesel production would introduce local jobs, and diverse production of carbon-neutral biodiesel fuels for energy and economic security.

Chapter Seven - Algae Culturing Techniques

Growth-Rate Calculus:

Calculating the growth of algae, follows the first-order equation: dCV/dt=uCV, where u is the "specific growth-rate" and CV is the "total cell-volume per Liter."

When you Integrate over time interval t1 and t2, the log-linear growth equation: lnCVt2 - lnCVt1 = u(T1-T2) is followed. Where ln CV is the natural log of the volume of cells per liter. If a cell culture is growing at a constant rate the plot of ln CV will be a straight line.

A Simple method for Calculating growth rates:

Algae, when introduced to a growth culture medium as an inoculant, start with an "acclimation phase" where growth-rates are initially inhibited. Algae cells become "shocked" when entering a new environment, and there is a period of acclimation which occurs sometimes for several days, to many days, with a new culture introduced into a new growth medium.

Algae growth, after an acclimation phase, enter an "exponential growth phase," where populations multiply rapidly, and with an increasing rate-of-growth. This exponential algal growth-phase is where algae researchers find their ideal conditions.

During this exponential algal growth-phase the "rate of increase" in cells per unit time, is proportional to the number of cells present at the beginning of the unit of time. The population-growth of the algae follows the equation: $dn/dt = rN$. The solution to this equation is the well known: $N(1) = N(0) e$ to the rt.

Measure the initial population of your algae $N(o)$, at the start time (T1), then measure your algae population $N(1)$ at the end of your time period. $N(t)$ - what you've produced - will equal $N(o)$, what you started with, having a growth rate (r) over the time period (t).

Once you measure $N(o)$, $N(1)$, over time period T solve for your growth-rate (r).

After the exponential growth phase available nutrients, or other factors of great interest to researchers are "limited," and growth rates suddenly slow and soon stop. If no new nutrients are supplied then algae cultures go into a rapid crash.

A biologist once said "biological systems, when stressed, either adapt, or die." This is very true with growing algae. Early algae pioneer pointed out: "Growth is limited, by that which it requires most" - Blackman (1905).

Algae Growth-rates are not the same as Algal Biomass accumulation.

Growth rates speak about number of cell divisions, and numbers of cells. Algal Biomass is concerned with total "mass" in terms of dry-mass of algae present at the start and end times.

Algae-"Yield" is determined by measuring the inoculant dry-mass in the beginning of your algae culture, and measuring the dry-mass at the end of your culturing period.
Balanced, and Unbalanced Growth in algal culturing is determined by the state-and-stage of algae growth occurring in your photobioreactor.

The specific Growth-rate is a "rate of change" of biomass and is determined by the magnitude of "anabolic" processes (photosynthesis), and "catabolic" processes (respiration): $U=P-R$ where U is the "specific growth rate" and P is photosynthesis and R is respiration.

The daily solar cycle of irradiance produces a daily "imbalance" of photosynthesis verses respiration. This ensures that "unbalanced" growth is the cornerstone of algal growth. This "unbalanced" growth is a great "trigger mechanism" in growing algae.

Algal species, are remarkable for their ability to "acclimate" to their environment. This characteristic is exploited by algal cultivators, by repeating conditions every day "training" the algae. Algal taxa respond with more predictable outcome.

Algae follows a traditional 5 phase growth cycle. These are acclimation, compensation point, exponential growth, saturation, then collapse (if nothing else is added). These 5 stages of growth follow a classic curve.

Acclimation occurs when you inoculate your growth media with a small amount of pure species. Compensation, occurs when photosynthesis exceeds the energies required by the cell for respiration, and reproduction.

Exponential growth, occurs next as all available algae consume the nutrients available. This phase is of the most interest to algal researchers. As maxima is reached, a saturation point occurs where the rate of growth diminishes. The final phase, is collapse.

As nutrients are depleted, micro algae cells begin to perish, and typically begin to sink. Without the constant production of lipids, cells are more dense, and sink.

Manipulating the cells, by limiting some variable (usually nutrient) you can "train" your algae to respond to different stimulus.

Early Work in Growing Algae

Algae Growing Pioneer, Otto Warburg (1931), won the Nobel prize for research, and explanation of Oxygenic Photosynthesis, describing respiratory pathways, using the green algae Chlorella. Warburg is a hero to the phycological field.

Growing algae, and micro-algae cultivation laboratory methods, finds its roots with techniques developed in the late 1800s, and early 1900s.

The early history of algae with mankind probably started when Paleolithic man harvested naturally occurring algae in tidal pools and ponds. Dried in the sun, algae would have added vital nutrients, and flavors to ancient recipes.

Growing algae in the modern-era began in the early 1500s, in Tokyo Bay, and continues to this day in Japan, and worldwide. Recent advances in algae growing methods have moved algae culturing (alga-culture) into rapidly growing markets for amino-acids, proteins, anti-oxidants, Omega 3 - rich lipids, and other organic molecules.

Algae is becoming the feedstock of choice for supplying nutraceutical, cosmetic, aquaculture, and biodiesel products.

Ferdinand Cohn (1850), the founding fathers of bacteriology, successfully kept and wrote about a unicellular flagellate form of Chlorophyae - Haematococcus pluvialis in his laboratory in Wroclaw, Poland. Haematococcus pluvialis is a valuable algae for its production of Astaxanthin.

Famintzin (around1871), St. Petersburg, Russia described his attempts at growing algae in a solution of a few dissolved inorganic salts.

Most algae growth occurs using a 12-hour on, and 12-hour off photoperiod. However, lengthening, or shortening the ratio has impacts on cell physiology and response. If "sun-hours" are increasing the alga know that summer is coming and increases photosynthetic response. If "sun-hours' are shortened alga respond to "coming winter" producing more lipids.

Culturing techniques include inoculating your growth media, measuring starting mass, setting photoperiod. Measure all Macro, and micro nutrients, metal ions, vitamins, as well as volume-mass-transfer of CO_2, and O_2 through your system.

Measuring your final mass, through Time T1-T2, you can calculate your Growth Rate.

Chapter Eight - Frequency Asked Questions About Photobioreactors

Question: What is a Photobioreactor?

A Photobioreactor (PBR) is a bioreactor stimulated by a light source(s). Usually this light source produces Photosynthetic Active Raditiaon (PAR) photon energy between 400 nm and 700 nm wavelengths. A basic photobioreactor includes optical growing vessels, aeration inputs, output vents, bacterial filters, Light Source(s), Lighting Timer, and Mechanical Structure.

Question: What are Algae Grow Kits?

Photobioreactor Algae Grow Kits as complete PBR hardware kits which you assemble. Algae PBR Grow Kits include a Mechanical Frame, Lighting System that produces a nominal 200 micro-moles/m2/sec of PAR light.

PBR Kits include a Heavy Duty timer and power system to control your photoperiod (usually 12 Hours-on/12 Hours-off) and fused power-plug. PBR Kits include a Pneumatic system of two (2) air-pumps, (4) check-valves and (4) biological filters (0.22 Microns) to remove bacteria from the aeration system before entering your Growth Vessels with four (4) Pyrex glass tubes for aeration into the growth vessels.

Question: Why Build a PBR Kit?

You can source your own materials, and build your own photobioreactor Kits. PBR kit will contain all of the basic lab equipment you need to grow algal taxa, in a controlled environment, with low Capital Cost.

Commercial grade PBRs, in the market, are typically expensive and offer many amenities and features such as data acquisition systems, not necessarily required, if you use "old-school" techniques such as titration tests.

Question: Can I Scale the PBR Kits?

Yes. PBR kits are scalable in capacity by simply adding more. Each kit has an 8 square feet footprint, and has a capacity of 80 Liters. To achieve higher capacities use multiple PBR kits. If you need 800 Liters of Algae-growing Capacity use 10 Kits.

Large Scale Example: (Note: PBR Kits are only for Indoor use, this example assumes an indoor appropriate work space)

One acre spans approximately 43,559 square feet. With room for aisles (70% packing one layer) between PBRs, you can install 3,812 Model X-80 PBR Kits for a production capacity of 304,960 Liters. Algal Biomass Harvest, with well-managed nutrient, water, air-quality and site operations can range depending on skill and species.

For example, (results will vary, but for purposes of illustration), a Chlorophyta can be harvested at 1 gram per Liter in well managed cultures. (Substantially higher densities are reported in the literature).

One gram/Liter/Growth-cycle would yield a gross (Dry-Weight) algal biomass of 304,960 grams (304 Kg)/acre/growth-cycle. Using 25 Days/Month at this rate yields, for example, 7,600 Kg per Month, (91,200 Kg/year) of Algal Biomass.

The Commercial viability of any large-scale Algal Cultivation System, requires special management and staff, adequate nutrients, water (and optional CO_2) inputs, and processing hardware for algal harvest, dewatering and drying. If you'd like to explore hardware costs at large scale please contact our offices.

Question: How much biomass can I grow with PBR Kits?

The English biologist Blackman, at the turn of the 20th century, said "photosynthesis is a process limited, by that which it requires most." Growth rates depend on how well you've balanced all factors including required nutrients (both macro and micro) dissolved ions, and vitamins.

The wavelengths and intensities of PAR light, with the photoperiod you set will influence your algae growth. The health of your inoculate when you start, and the mass-transfer of CO_2 from the atmosphere during growth (aeration during cell respiration) in the form of dissolved CO_2 and O_2, as well as the pH of the growth media management throughout the growth-cycle will also dictate your growth results.

Algal biomass (dry weight) growth of 1 gram/Liter, per cycle is repeatable, but can range lower or higher depending on your skill, taxa, and balance of systems such as temperature, pH and nutrient mix chosen. Algal dry-weight yields from PBR use have been reported from 5 to 10 grams per liter. Your results depend on your growth media, taxa, PAR light, photoperiod, and skill. You can achieve a repeatable 3-4 grams/liter with this hardware.

Question: How much Light does the PBR Kit Produce?

Photobioreactor (PBR) Kits include two (2) high-efficiency ballasted T8 Florescent Lighting Fixtures. Four (4) high-efficiency T8 bulbs are included at 6500K spectral output. You can replace the bulbs with different spectral profiles easily using the T8 size. Nominal output at height is 200 micro-moles photons/m2/second of PAR light which you can adjust up or down by using different vertical leg segments, or by suspending the light at different heights with chain suspension included. Bulbs are rated for 20,000 hours of use.

Question: How long does it take to assemble the PBR Kits?

PBR Kits are easy to assemble and relatively fast. Assembly of the complete Kit takes about two hours if you're slow and steady. Note: when you are ready to inoculate dis-assemble the growing vessel connections and use the included sanitizer (100% Non-Toxic) following included instructions, which evaporates leaving your working surfaces ready for quick connection, and you're ready to inoculate your starter algal taxa.

Question: What's included with the Pneumatic system in PBR Kit?

PBR kits include a high-efficiency air pumping system composed of two (2) air pumps, (4) check-valves, (4) 0.22 Micron Bacterial Filters (one for each Glass Growing Vessel), with 100% non-toxic Food-Grade plastic tubing (22') and fittings and four (4) Pyrex glass tubes for aeration into your algal growth vessels, as per the Parts list in Chapter Three.

Question: How do I control Temperature?

These photobioreactor algae PBR kits are designed for indoor use. To control the temperature of your photobioreactor growth vessels you can control the environmental temperature of your lab space, or you can add heating elements such as hot plates you can source locally. Most algae grow at temperate levels around 20 degrees C.

Question: How do I Harvest Photobioreactors?

Each glass growing vessel, 20 L, or 25 Liter Size, (Kit contains Four vessels) comes equipped with a quick release full seal Top Stopper. (Use 100% non-toxic food grade plastic). When you wish to access your growth vessels, either for loading growth media, taking samples, or harvesting, remove the Top Stopper, and insert your Pipette, or other glassware to pump or manually extract your samples or harvest. Replace Top Stopper when you've finished your extraction. Do not turn-off air pumps. They should run 24/7.

Question: How do I Stir Cultures?

The Mechanical Framing included in the PBR kit design, allows easy access to all components. Even though there is a gentle mixing pneumatically from the air pumps, you can easily "Swish" the vessels manually giving the alga a gentle but good stir without opening the vessels.

Question: Do I need Special Tools to Assemble the PBR Kits?

No. Sharp Edge, Measuring Tape, Scissors and plastic gloves (recommended). Once you've assembled your frame you can glue the parts with PVC glue locally sourced.

Chapter Nine - Quick-Guide to Photobioreactor Construction

Photobioreactor algae grow kits are designed for researchers who want to conduct experiments, and need the hardware to cultivate algal monocultures.

Use photobioreactor PBR algae grow kits to create controlled photosynthesis and farm algae for its incredible and valuable proteins, amino-acids, lipids, anti-oxidants, vitamins and other amazing compounds
Photobioreactor PBR Algae Grow Kit - 80 Liters for growing monocultures of algae.
resent in your source water.

Step One: Assemble Frame of PVC piping you source at local hardware. Cut lengths as described in Chapter Three.

Step Two: Assemble Glass Growing vessels, with 2-hole stopper (100% food grade non-toxic plastic). Through one hole, slide Glass Pipe (4 mm) nearly to the bottom of the glass vessel with 2 inches extending above the stopper. This is your air intake glass pipe. To the top of the other hole, insert a Pasture's Curve extending up from the stopper base. This is the "exit" valve allowing internal air pressure to build up, and provide a constant out pressure.

Pasture's curve prevents bacteria from creeping back into the vessel.

Step Three: Assemble the air pumps. You'll use two air pumps, from an aquarium supply house, with a splitter and two "Check Valves." You'll pump air into two Growth Vessels with one pump. From each pump, place inline one Check Valve, and before each vessel, you'll place a 0.22 um Bacterial Filter. This will remove any bacteria, or particulates from the incoming air.

Step Four: Connect, using 100% non-toxic Food Grade tubing, the Bacterial Filter to the Air Intake pipe in Hole One of the Stopper. Length of plastic tubing approximately 22." Air, is now being pumped from a pump, through a splitter to run to Growth Vessels. Each "leg" from the pump "splitter" will have a check-valve, and a Bacterial Filter. With the tubing, as described, connect the down-stream side of your Bacterial Filter to the air Intake pipe at Hole One on the stopper.

Step Five: Assemble the Florescent Lighting fixtures, and place them on top of the Mechanical Frame. Plug in lighting units to a Power Strip. Plug the Power strip into a Timer, and plug the timer into the wall current.

Step Six: Dis-assemble tubes, and glass ware, and soak in sterilizer (evaporative type), before you load the vessels with Growth Media, and Inoculate.

There you have it, a photobioreactor you can build yourself. Grow algae for profit, by growing highly valued species.

CPSIA information can be obtained
at www.ICGtesting.com
Printed in the USA
FSOW02n1302300517
34802FS